PILLY

the Little pill bug with Big secrets

Written by Monterrey Williams
Illustrated by Laura Osborn

To our wonderful, lovely Juniper - who is as cute as a bug!

Love,
Grandma & Papi

ACKNOWLEDGMENTS

I want to thank Laura for her dedication and interpretive talent to make the story come alive. Also, I want to extend a special appreciation to Toni Winkler for editing.

ISBN:1515151190
ISBN-13: 978-1515151197

DEDICATION

To Jake and Beau

HOME SWEET HOME

My name is Pilly and I'm a pill bug. My home is under a log in the green forest. I live with my Mother and Father and all my brothers and sisters. It is a great place to live. I have plenty of yummy food to eat. My log house serves up a delicious buffet of leafy lichens, gray moss, and green algae. I also like to munch bark too. Grandmother lives at the other end of the log. I visit her twice a week. Today is a special day, since I am two years old and Grandmother has promised to tell me amazing secrets. Mother always goes with me to Grandmother's house, however, today, she told me to go all by myself.

The little pill bug started down the crusty log to Grandmother's house. As he climbed through a crevice in the bark, he saw his friend, Lucky the lady bug, sitting on a tiny stem. "Hi Cuz!" Lucky called out with affection. She called him Cuz because they are both bugs. "Where are you walking this morning?" inquired Lucky. "Don't forget, it's the Fourth of July," she added. "Hi Lucky," Pilly smiled. "I'm going to visit Grandmother. She's going to tell me some secrets today." "Oh no kidding? You're going all by yourself?" Lucky asked, looking at Pilly with interest. "What kind of secrets?" she asked. "Secrets to make me proud of myself," answered Pilly. "That's what Grandmother told me. I'm two years old today and it's time for me to learn the truth about myself," he said. "Ok, I'll be waiting for you when you get back Cuz. We'll watch the fireworks together." "That will be fun," Pilly told her. He shivered with excitement. He couldn't wait until tonight.

Pilly walked a short distance and saw the lichen patch up ahead. He knew he would have to be careful crawling over the leafy edges; not to trip. Having seven pair of legs was a challenge at times because all of his feet needed to be moving at the same time and in the same direction. Stopping in the middle of the green lichen patch, he couldn't resist munching on the sweet curled leaves. This was his favorite food. A similar lichen grew near the green one, but it was a brown color and looked like dried paint. However, Pilly preferred the leafy green one. "Well, I better not dally too long," he said to himself. "Grandmother is waiting for me and I can't wait to hear the secrets."

It was a beautiful day. The dappled sunlight, filtering through the leaves on the trees, made the log look enchanted. There were various shades of yellows, greens and grays with splashes of pink here and there. He was fascinated and loved the kaleidoscope of different colors. To the left of his log house was a field with native flowers and wild grasses. Still wet from the morning dew, hundreds of tiny spider webs connecting the blades of grass, were made visible. The silky strings glistened under the sun's rays. Dew drops hung heavily on the webs and it looked as if they had been sprinkled with a thousand dazzling diamonds. What a spectacular sight, Pilly thought, and a perfect way to start the Fourth of July.

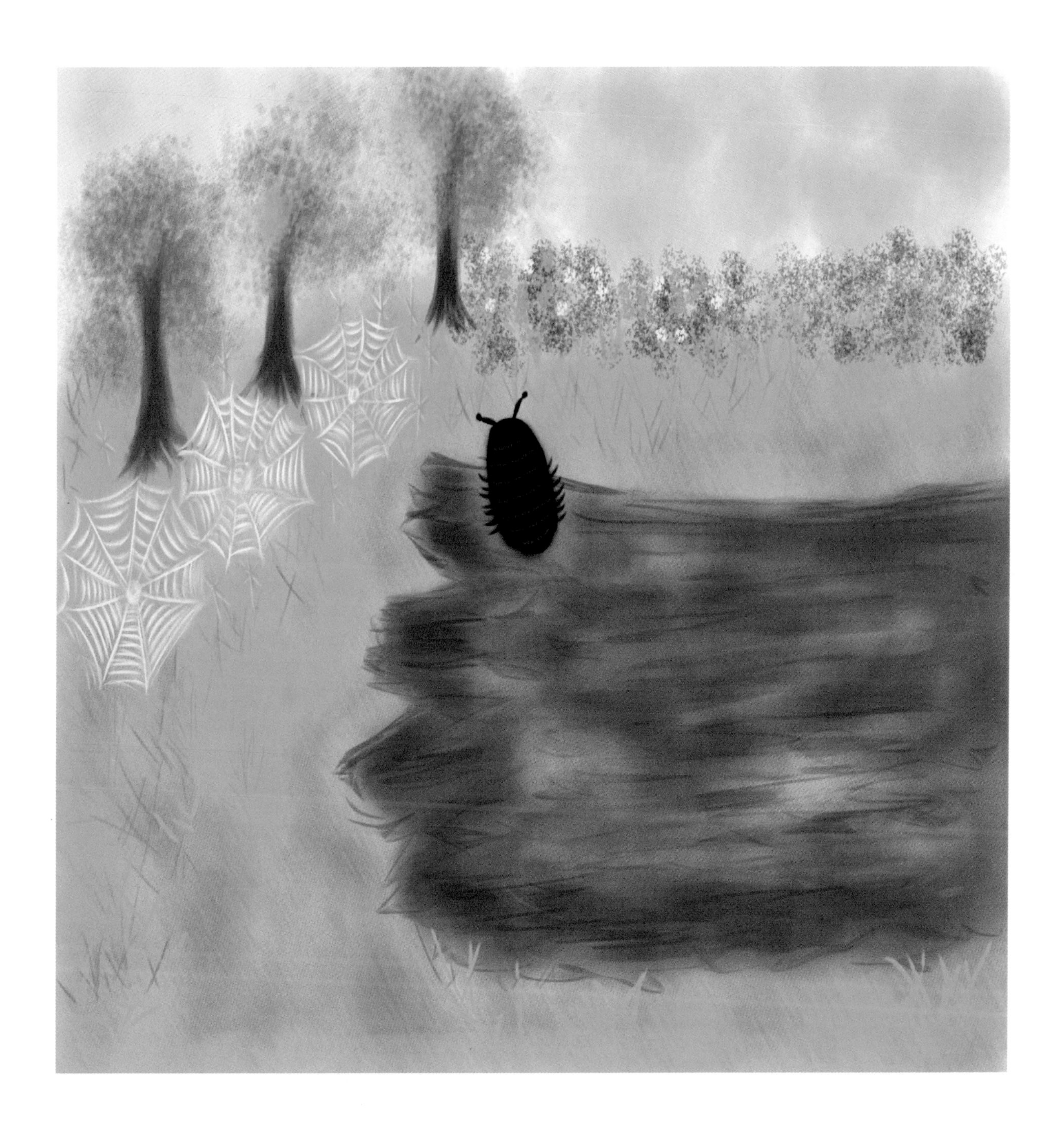

In the middle of the grassy field was a pile of rocks and small boulders. Many times Pilly ventured over to the rock pile to see what he could find. As he traveled over to the rocks, he would play a game of you can't touch the ground. He climbed on twigs, dandelion stems, acorns, leaves, pine straw and vines--anything that made a bridge and take him closer to the rocky pile. Sometimes he had to climb to the top of a blade of grass before it would sway, bend, and finally topple down to the next platform. It was a fun challenge and most days he made it all the way over to the rock pile without touching the ground.

Suddenly, distracted by two comedians, Pilly laughed out loud. Two squirrels were chasing one another round and round a tree. As they jumped off the tree, they both hit the ground at the same time and bounded to the next tree doing the same thing. Finally, they tired of the game and clawed at the ground to find some acorns.

He walked a little further and saw a bunch of Spanish moss on the log. It looked like a big gray tumbleweed. Any other day, he might be tempted to sit and relax on one of the windows inside of the matrix. Not today though, since he was in a hurry to get to Grandmother's house. Instead he climbed over the moss and continued his way down the old vine attached to the log.

Oh no! Here comes Shadow, the frisky dog that lives in the house at the edge of the forest. Shadow put his big black nose next to the log and looked and sniffed and looked and sniffed. "He is such a curious dog," thought Pilly, as he stared in the big brown eyes. Soon, the dog bored with looking and sniffing and just about that time, he heard someone calling his name from the house. He was gone in a flash. "Finally," Pilly said with relief, I had to stay frozen in one place, while Shadow was here. I didn't want Shadow to think I was something to eat! Now, I can get to Grandmother's house.

The vine on the log was super highway and he made good time following it. Before long, the vine curved down and went on the side of the log. He finally reached the other end of the log. From this point, he followed a well laid ant trail to the entrance of Grandmother's house.

Carefully, he made his way into the interior part of the log. The log had been hallowed out and it made a grand entrance to her house. "I think I see Grandmother," Pilly told himself. Sure enough, Grandmother was sitting by a tiny pool of water in the middle of a leaf. "Pilly! My dear little roly poly!" Grandmother always said that to him. It was her "love" name for him. "I've been waiting for you," Grandmother said. "Come rest those legs of yours and sit by the pool with me." Pilly walked up to the leaf pool and got a cool drink of water. Nothing tasted more refreshing; especially after walking all the way across the log to the other side from his house.

"Grandmother, do you know what today is?" he asked. "How could I ever forget!" Grandmother exclaimed; with a twinkle in her eye. "How was your walk over here?" she asked. "It was great," Pilly said. "I got to say hi to my cousin, Lucky the lady bug. Then I stopped to get a snack from my favorite food." "Hmmmmm, I wonder what that could have been?" she smiled; knowing he was crazy about lichen. Pilly giggled. "I wanted to play in the moss for a while, but I didn't want to dally. I would have been here sooner but Shadow, that pesty dog that lives in the house by the forest, came over to the log. I had to stay as flat and still as I could because I didn't want Shadow to see me," Pilly explained.

"My, my, what an adventure! Well now...I am so happy you are here today because I'm tired of keeping secrets from you." Pilly's eyes widened, as he smiled, with anticipation. "So...here is the first secret I wanted to tell you today, my love," Grandmother said. "There are some pill bugs that are called *runners* because when they get in a scary situation, they just run. But you, Pilly , were born with a special ability," Grandmother smiled tenderly. "You can roll up in a ball like an armadillo! So, if you are frightened, or the next time Shadow comes over to your house for a visit, you have the option to stay, run or roll up in a ball." "Oh! Wow, that's pretty cool," Pilly said; already thinking where and when he could use this trick. "When can I start practicing?" he asked. "Dear, you've always been able to do it. I just wanted to be the first pill bug to tell you about it."

So, right then and there, Pilly flexed the plates on his back, stretched a bit and Pop! He rolled himself into a ball. "Look Grandmother! I'm a little armadillo!" "Good job," she said proudly. Her Grandson was smart and learned the trick quickly.

"I have another secret, my dear, that I want to discuss. It is a miracle of sorts and it has to do with where you were born." she explained. "Grandmother, I already know I was born in Texas," Pilly said quickly. Grandmother laughed out loud. "Honey, I'm not talking about Texas! I'm talking about how your Mother held you in a little pouch, like a kangaroo does her babies!" "Wow," said Pilly. "I guess I was really loved and protected." "You were," she continued. "Your Mother always wanted a big family and you were one of about a hundred. She loved and cared for every one of you; in this special way. Would you like to have a piece of lichen before you leave today? There's a wonderful patch of it right over there." "No, not today. I'm so excited to learn about the secrets, I can't wait to tell cousin Lucky!"

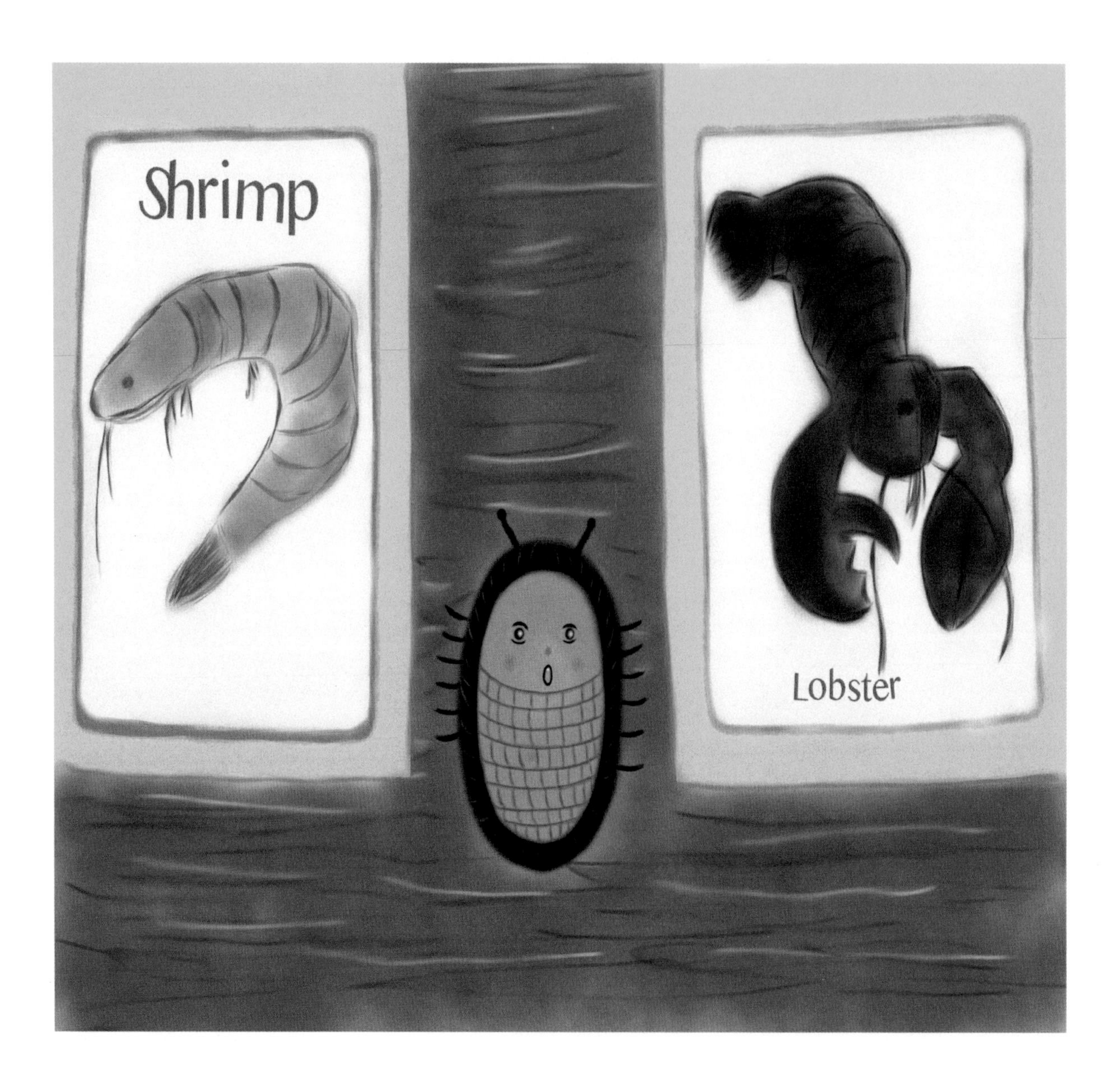
Shrimp
Lobster

"That reminds me, Pilly," Grandmother hesitated for a moment; wondering if she should continue. "Hmmmm, there is one last secret to tell you. This is the biggest secret of all, since hardly anyone knows about it. We...pill bugs.....are...crustaceans," she said carefully; watching the reaction on his face. "We are related to shrimp and lobsters." "Wow," Pilly said with surprise. "I'm not a BUG? I'm a CRUSTACEAN?" "Yes, dear, we have gills with which to breathe! But be careful! That DOESN'T mean you can only live in water."

"Double wow!" Pilly exclaimed. "Now I really can't wait to see Lucky. I want to show him the cool trick of rolling into a ball. The next time I see Mother, I'm going to give her a big hug for loving and taking such good care of me. But, Grandmother, can we keep it a secret that we're not bugs?", asked Pilly. "I like being called Cuz," he said concerned. Grandmother smiled, understanding. "That's fine Pilly. I want you to know our family comes from a fine line of Crustaceans! It's something of which to be proud." "Wow, it's the Fourth of July today," said Pilly, "but I feel like celebrating before the fireworks begin. Grandmother, may I still have a piece of lichen? This call for a celebration right now!"

THE END

Made in the
USA
Monee, IL

15461644R00021